STEM Middle School

Building Blocks of Engineering Student Workbook

G. Grant

Table of Contents

Letter to the Student

Dear student,

If you are opening this book, you are already well on the way to an enriching and successful future. The key to achievement in life is to always seek more- more knowledge, more skills, and more resources. I am honored to share with you some innovative and exciting educational tools that can help you develop not only your scientific knowledge but also your ability to manipulate that knowledge in order to accomplish new and great things.

STEM- science, technology, engineering and mathematics is the new approach towards integrating real-life problem solving into your science and math classes, leaving you with skills that you can actually use throughout your life.

Let's start with the science. Science is how we process and understand the world around us. Science is meant to be built, touched, and experienced. As you embark on this journey to learn and experience science, remember the great men and women before you - they didn't simply memorize or imagine abstract concepts- science was real to them and that is what allowed them to understand it. This course ensures you get that chance to fully experience the science- hands and mind fully engaged.

Of course we see how great a role technology plays in our lives. Work, medicine, transportation, play- is there an area where we don't see some aspect of a technological advance? If we want to stay current and keep ourselves relevant it is important that we understand how technology works- only then can we devise new ways that technology can help our society. We all know that just because someone knows how to text it doesn't mean they can build a touchscreen. Well, after this course- you can.

What does it mean to be an engineer? Simply put, engineering is problem solving. The ability to assess a situation, weigh the pros and cons, allocate your resources, and search for an innovative solution reaches well past your school years. The critical thinking and problem solving skills you acquire from this course will accompany you throughout life, and you will confidently come to rely on them as you face life's challenges. Never under estimate your ability to solve a problem- any problem.
Whether you become a professional athlete, teacher, architect, stay at home parent or any other vocation, your social skills and ability to navigate in a group will serve you well. You will learn that life is not about sitting alone at a desk and taking your own notes- real life requires communication and collaboration with others- like minded and oppositional. Through the structured activities in this STEM course, you will learn when to lead, when to negotiate, how to merge different ideas, and how to be a team player.

Talking about math with students is always scary - nobody wants to admit how important the numbers really are. But they are important. You want to build a bridge? Angles matter. You want to figure out how fast the roller coaster can go without flying off the track? Math again. Instead of seeing math as the enemy, an incomprehensible set of rules for how numbers should and should not be combined- this course will teach you simple understandable ways to use math. Physics is the physical manifestation of math- and it is going to make it so simple that you are going to be figuring out the formulas yourself- trust me.

I wrote this course to empower you as a student, as an engineer, as a collaborator, as a problem-solver, as a critical-thinker, and as a future leader. Our society needs you. We need you to be resourceful, knowledgeable, and able to innovate.

So learn, grow, and give back.
G. Grant

STEM Advancement Inc.

131 Bloomingdale Rd

Staten Island, NY 10309

(917)-838-7864

Email: Stemmiddleschool@aol.com

Website: advanceSTEM.com

ISBN-13: 978-1540819437

ISBN-10: 1540819434

Introduction to Engineering

The Essence of engineering can be stated very simply: Problem Solving.

The Situation:

A few friends are gathered after school to work on a science fair project. It's dark outside when Jeremy cuts his finger open with the razor blade meant to cut shapes out of the model airplane. You quickly put some pressure on the wound and call the paramedics; thankfully, they are on their way. However, five minutes later you realize that since it is dark outside, even though you told them that you live at 11 Elm Street, the paramedics will have no way of knowing which house is yours! Use the materials below to construct an emergency sign to help the paramedics find you. Hurry!

The materials:

2-4 plastic straws

2 three volt batteries

2-4 colored LEDs

The solution:

STEM Middle School

Materials Engineering Lesson # 2

Main Branches of Engineering

The Essence of engineering can be stated very simply: Problem Solving.

Houston-we have a problem!!

SOS! The Amazing Adventure Park has called down your team of engineers to help them- a group of kids got stuck upside down on the roller coaster!! Below is a list of tasks necessary for saving the kids and fixing the ride.

Sort:

1. creating a safe-landing pad for the kids in case they fall
2. making sure the landing pad is waterproof and nonflammable
3. building a ladder tall enough to reach the kids
4. adding wheels to the ladder in order to roll it under the roller coaster
5. rewiring the computer that controls the ride
6. making sure none of the wheels on the ride are twisted or off the track
7. developing a medicine to rid the kids of nausea once they come off
8. making sure the landing pad has a hole in the middle for the ladder
9. making sure the ladder doesn't look scary and looks fun for the kids
10. creating a tool that can rip through the seatbelts easily to release the kids

Your job is to correctly assign the tasks to each of the following people:

1. civil engineer	2. electrical engineer	3.chemical engineer
4. mechanical engineer	5. systems engineer	6. designer

Appeal vs Function

An engineer and a designer work hand in hand to create a marketable product.

Define:

____________ is an engineer's job.

____________ just improves the ability to market the item, but does not make it any more useful.

Classify:

Which of the following is real engineering?

1. Designing a bridge so it can hold up the weight of 200 cars.
2. Designing the structure of the bridge's crossbeams.
3. Designing the shape and color of the bridge's crossbeams.
4. Making a radio that is waterproof.
5. Making a radio that can glow in the dark.
6. Making a radio that can project light onto the ceiling.
7. Making a radio that can turn different colors.
8. Making a radio that is square in shape.

Building Blocks of Engineering

Electronic Cord Holder Design and Discovery Lab

●	Design opportunity – **Build a device that can effectively hold several electrical cords together.**
●	Research- study different devices meant to hold things together, paper clips, hair clips, etc.…
●	Brainstorm- write all your ideas below
●	Sketch out your best idea below
●	Materials- list your materials below, and plan a procedure
●	Build it
●	Evaluate your design and how well it fulfilled the design opportunity
Additional Info	

Building Blocks of Engineering

Scamper

 SCAMPER is an acronym as well as a technique. When you Scamper, you do not invent a new item; rather, you improve an existing one.

Plan:

Please SCAMPER a school bus.

S

C

A

M

P

E

R

Building Blocks of Engineering

Tension and Compression Forces

 A force can do different things to an object, depending on how the force is applied.

Define:

A force is a _____ or a _____.

Tension is _____

Compression is _____

Compare:

Please fill in the charts with High or Low to indicate the various forces that can act on the material without breaking it. For example, concrete can withstand a lot of force pressing inwards and is high in compression, but it cannot be stretched even a little and is therefore very low in tension.

	Tension	Compression
Concrete	Very low	high
Wood		
Plastic		
Brick		
Steel		
Reinforced Concrete		

Stress vs Strain

When you are under a lot of stress, you feel very tense!

Classify:

Is this stress or strain?

1. Placing your foot on a beach ball_____

2. The beach ball deflates _____

3. The bridge collapses _____

4. The elephants crowd onto the bridge _____

5. The baby's suspenders snap open _____

6. The baby's suspenders are stretched two inches more than normal _____

7. What kind of reaction is this?

8. A wooden block holds up three people with no visible strain _____

9. A glass window shatters when the baseball hits its center _____

10. A plastic ruler is bent and snaps back to its original shape _____

11. A plastic rule is bent and snaps back to its original shape, but it has fold lines

    _____

12. A metal drawer is kicked hard _____

13. The metal drawer is now dented permanently _____

Building Blocks of Engineering

Tensile Lab

 Tension is the force that stretches or pulls outwards.

Experiment:

Type of material	Textbooks (or pounds) until material deformed	Textbooks (or pounds) until material fractured
paper		
aluminum		
Cardboard		

Analyze:

1. Which material was able to withstand the most tension?

2. Would the results be different if this were a compression test? Why?

3. Look carefully. Some materials fractured or just pulled out of the clamps at what we call "the weakest point." Explain.

STEM Middle School
Materials Engineering Lesson # 10

Compression Lab

Hardness can refer to several different measurements. In this lab we are referring to the ability to resist a compression force.

Experiment:

Type of cup	Textbooks (or pounds) until cup deformed	Textbooks (or pounds) until cup fractured
Regular plastic		
Styrofoam		
Hard plastic "tumbler"		
Block of wood		

Analyze:

1. Which material was able to withstand the most compression? Which is the hardest?

2. Would the results be different if this were a tension test? Why?

3. Look carefully. Did all the materials fracture, or did some just deform very badly? If a material didn't fracture, what could be a reason why?

Density Lab

 The density of water is one gram per cubic centimeter. This means that anything with a density that is greater than one will sink, and anything with a density of less than one will float.

Experiment

Item	Mass	Volume	Density	Predict: Will it float or Sink?	Was your hypothesis correct?	What materials are like this one?

Remember: Volume of a cube = length x width x height

Do your calculations below.

STEM Middle School
Materials Engineering Lesson # 12-13

Properties of Materials

 The properties of a material are determined by its internal structure. Knowledge of internal structure allows you to understand and use materials appropriately.

Record:

Fill in the chart with a number from 0-10. 0 means that property does not apply to it at all, 10 meaning that property describes it perfectly.

Material	Ductile	Hard	Electrical/Thermal Conductor	Melting Point
Metal				
Ceramic				
Plastic				
Organic				

Compile:

Write all the properties you can about this material

Metals	
Ceramics	
Plastics	
Organics	

What is a Newton Lab

 Sir Isaac Newton was a mathematician and physicist who lived in the 18th century.

Experiment:

Fill out the chart below with your measurements. Make sure to include the units!

Weight	Newtons Required to lift it TRIAL 1	Newtons Required to lift it TRIAL 2	Newtons Required to lift it TRIAL3	Newtons Required to lift it AVERAGE
500 g				
450 g				
400 g				
350 g				
300 g				
250 g				
200 g				
150 g				
50 g				

Extension:

Create a graph correlating to the information you gathered

Conclude:

How many grams can be lifted with a force of 1 N?

Building Blocks of Engineering *(vertical text, left margin)*

Build Your Own Spring Scale Design and Discovery Lab

●	Design opportunity – **Build a device that can measure the force used to lift an object.**
●	Research- study the materials provided by your teacher 　　　- one long yardstick or ruler　　-long paper strips to cover the ruler 　　　-tape-rubber band 　　　-paper clip　　　　　　　　-50 g weights
●	Brainstorm- write all your ideas below
●	Sketch out your best idea below
●	Materials- list your materials below, and plan a procedure
●	Build it
●	Evaluate your design and how well it fulfilled the design opportunity
Additional Info	

Lab

The triangle is a very structurally sound shape, as any force placed on a point is immediately distributed down the sides, which have a much larger area.

Define:

1. Structure _____

2. Load _____

3. Force Distribution _____

Sketch:

1. Draw a triangle and show how a compression force is distributed.

2. Draw a square and show how a compression force is distributed.

Compare:

Compare and contrast the relative stability of a square and a triangle. Which is a more stable structure? Why?

Summarize:

How can you make a square more stable?

Suspension and Truss Bridges

 A covered bridge made out of wood can last over 200 years!

Compare and Contrast:

	Suspension Bridge	Truss Bridge
High or Low in Compression		
High or Low in Tension		
Triangular Shape		
Uses Rope		
Used for massive loads		
Exists Nowadays		
Able to be built by hand		

Refute:

Josh thinks that it is always preferable to build a suspension bridge to a truss bridge. Can you refute his claim?

Building Blocks of Engineering

STEM Middle School
Materials Engineering Lesson # 18

Spaghetti Tensile Test

 Tension is the pulling or stretching force.

Experiment:

Strands of spaghetti	How many Newtons until deformation	How many Newtons until fracture
One		
Two		
Three		

Record:

Write down the procedure you used to measure the tensile strength of a piece of spaghetti.

Improve:

What can you do to increase the tensile strength of the spaghetti?

What can you do to ensure more accuracy and precision in this experiment?

Build a Popsicle Stick Bridge Design and Discovery Lab

●	**Design opportunity** – **build a bridge of any design using only 20 popsicle sticks and glue. The bridge must span 14 inches and will be tested to see which can bear the most weight.**
●	Research- study the criteria and determine if the bridge should be built to withstand a compression force or a tension force.
●	Brainstorm- write all your ideas below
●	Sketch out your best idea below
●	Materials- list your materials below, and plan a procedure
●	Build it
●	Evaluate your design and how well it suited the design opportunity
Additional Info	

Building Blocks of Engineering

Conductors and Insulators

 A test circuit is an open circuit with an indicator built into it. An insulator will not allow charge to flow through it, and a conductor will allow charge to flow through it.

Experiment

Build a test circuit and try to complete it with each object in the bag that your teacher provides. Record whether the objects are conductors or insulators on the chart below.

Conductors	Insulators

Infer:

What do you think a WIRE is made of?

Building Blocks of Engineering

STEM Middle School
Materials Engineering Lesson # 21

Optical Properties

 Light is a kind of energy called electromagnetic radiation. Light is both a wave and a particle.

Experiment:

What happens to the beam of light when you shine it through the following objects?

Object	What happens to the beam of light?
Clear plastic divider	
Colored plastic divider	
Hard notebook	
Mirror or silver foil	
Cup of water	

Infer:

What do you think would happen if you aimed a beam of light through smoke?

Building Blocks of Engineering

Forces

A force is a push or a pull.

Matching

1. force   a. a force in a certain direction

2. unbalanced force   b. 2 forces of equal and opposite magnitude

3. vector   c. Newton

4. balanced forces   d. net force after 2 or more forces are applied

5. resultant   e. a force in a certain direction

6. unit used to measure force   f. a force that results in motion

Draw the resultant vector:

1.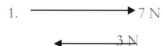

2.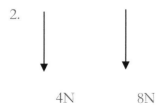

Forces continued

A force is a push or a pull.

3.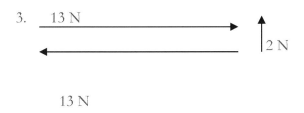

    13 N

4.

    14 N

    6 N

5. 4 N

    2 N

Building Blocks of Engineering

Forces Lab

 Balanced forces have equal magnitude in opposite directions

Experiment:

Use your spring scale to test the scenarios below. You will need a second spring scale to use as the other unbalanced force. Fill in the chart as you go along.

What is being measured?	How much force does it require?	What is the resultant?
Push a penny across the desk		
Push two pennies across the desk		
Have a friend use only 1 Newton to stop the penny from moving. Try to push it again.		
Have a friend use only 1 Newton to stop two pennies from moving. Try to push it again.		
Hook a rubber band between 2 spring scales. Have a friend use 0 N. to hold back the rubber band, try and stretch the rubber band		
Hook a rubber band between 2 spring scales. Have a friend use 5 N. to hold back the rubber band, try and stretch the rubber band, use only 5 N.		

Building Blocks of Engineering

Forces Lab Continued

 Balanced forces have equal magnitude in opposite directions

Hook a rubber band between 2 spring scales. Have a friend use 5 N. to hold back the rubber band, try and stretch the rubber band, use as much force as necessary to fully stretch the band.		

Conclude:

In which of the above examples is the resultant the same as the force required? Why is this true?

STEM Middle School

Basic Electrical Engineering Lesson # 3

Effect of Distance on Magnitude of Force

 Magnets are generally a safe and inexpensive way to experiment with forces.

Problem:

What is the effect of distance on the force acting between two magnets?

Hypothesis:

Materials:

Ruler

2 magnets

String

Notebook

Challenge:

Find a way to quantify how close together magnets can be to attract or repel each other using only the materials listed above.

Building Blocks of Engineering

STEM Middle School
Basic Electrical Engineering Lesson # 4

Effect of Amount of Particles on Magnitude of Force

 Magnets are generally a safe and inexpensive way to experiment with forces.

Challenge:

Mary thinks "the fewer magnets there are, the stronger the magnetic force between them".

Use a pencil, a ruler and five circular ceramic magnets to prove Mary correct or incorrect.

Hint: The magnets can fit over the pencil.

STEM Middle School

Basic Electrical Engineering Lesson # 5

Energy

Energy is the ability to do work. Hope you have the energy to do this worksheet correctly!

Classify:

Which type(s) of energy...

1. are responsible for keeping protons (which should repel each other) in close proximity?

2. makes solar powered flash lights work at night?

3. makes bicycles move?

4. make electric scooters move?

5. are what allows an electric kettle to boil water?

6. makes people wear sunglasses?

7. causes the walls to vibrate at a party?

8. makes slamming a door shake the house?

9. Is what makes your toast burn if you leave it in too long?

10. makes your microwave work?

11. do FIOS customers make use of?

12. is used to power windmills?

13. is used in many power plants?

14. holds hydrogen and oxygen together in your cup of water?

15. produces electricity in a battery?

Building Blocks of Engineering

Energy Transfers

 Energy cannot be created or destroyed, only converted into another form.

Diagram:

Use arrows to show which energy transfers occur in the following scenarios.

1. An alarm clock suddenly starts blaring and shows that it is 7:12 am.

2. You eat a huge dinner and make a super layup at the basketball game.

3. The bread you put in the oven springs up with a ding, the red light on the toaster and charred bread indicate that your toast is ready.

4. a nuclear reaction takes place inside a chamber, the chamber grows very hot. Steel rods are inserted into the chamber, and then placed in cold water. The cold water immediately turns to steam and the steam turns a generator which produces electrical energy.

5. can you write your own scenario here?

Building Blocks of Engineering (vertical, left margin)

Energy Transfer Design and Discovery Lab

●	Design opportunity – **build an object that can transfer one type of energy to another**
●	Research- look up possible energy generators and sources and converters
●	Brainstorm- write all your ideas below
●	Sketch out your best idea below
●	Materials- list your materials below, and plan a procedure
●	Build it
●	Evaluate your design and how well it suited the design opportunity
Additional Info	

Building Blocks of Engineering

Charging a Neutral Object

Matter is inherently neutral-it has equal numbers of positive protons and negative electrons.

Evaluate:

Find the charge on the atom…

1. 4 protons, 4 electrons _____
2. 4 protons, 3 electrons _____
3. 4 protons, 2 electrons _____
4. 4 protons, 1 electron _____
5. 5 protons, 4 electrons _____
6. 5 protons, 3 electrons _____
7. 18 protons, 12 electrons _____
8. 12 protons, 18 electrons _____
9. when opposite charges pull together, we say that they _____
10. when like charges push apart, we say that they _____

Classify:

In which way is this neutral object getting charged?

1. sock rolling around in the dryer
2. your iPhone getting plugged into the computer
3. your plug is placed right near the outlet and receiving shocks
4. a balloon being rubbed on your hair
5. clouds rolling by each other in the sky

STEM Middle School
Basic Electrical Engineering Lesson # 9

Static Electricity

Your teacher will set up several stations with different static electricity experiment. Visit each station and answer the questions below.

Station 1: Bad Hair Day

Materials: Balloons and your hair

Procedure: rub a balloon vigorously against your hair for at least 30 seconds. Try and stick the balloon to a wall. If this does not work, rub more vigorously for a longer period of time.

Questions:

What happens to electrons as you rub the balloon against your hair?

Why does the hair seem to be attracted to the balloon?

Why does the balloon stick to the wall?

Station 2: Balloons and soda cans

Materials: Balloons and an empty soda can

Procedure: rub a balloon vigorously against your hair for at least 30 seconds. Place the balloon close to a soda can and watch the soda can roll away!

Questions:

What happens to electrons as you rub the balloon against your hair?

What charge do metals normally have?

Why does the metal move away from the balloon?

Infer: When you rub a balloon against your hair, which one is losing electrons and which is gaining? How do you know based on the results of this experiment?

Static Electricity Continued

 Your teacher will set up several stations with different static electricity experiment. Visit each station and answer the questions below.

Station 3: Lightning Sparks

Materials: balloons, metal, dark room

Procedure: rub a balloon vigorously against your hair for at about 2 minutes. Close the lights and touch a piece of metal-you should see a spark!

What is that spark made of?

How is the spark you saw similar to real lightning?

Name the process that is occurring when you see the spark.

Building Blocks of Engineering

Static Electricity

 In today's lesson you will be using a spring scale to drag an object over several different surfaces. You will measure the force required to drag the object across each surface. Fill in the sheet as you go along.

Order:

Place these surface materials in order, from causing the most friction to causing the least friction. This is a hypothesis, as you will not know the correct answer for certain until after the experiment!

silver foil

lab table

silver foil with a few drops of oil smeared on it

thick fabric (such as a sweatshirt)

crumpled paper

Experiment:

Attach a 50 g weight to the end of your spring scale. Drag it over each surface and determine how much force is required to drag it. You will be comparing each amount of force with your control, the amount of force it takes to drag 50 g across the lab table. Write your results below.

Surface	How much force	More or less friction than control?
Lab Table		This is the control
Thick fabric		
Crumpled paper		
Silver foil		
Silver foil smeared with oil		

Static Electricity Continued

In today's lesson you will be using a spring scale to drag an object over several different surfaces. You will measure the force required to drag the object across each surface. Fill in the sheet as you go along.

Evaluate:

Evaluate your hypothesis (question 1). Where you correct or incorrect? What did you learn?

Extrapolate:

What would happen if you would add wheels to the 50 g weight?

STEM Middle School
Basic Electrical Engineering Lesson # 11

Building Blocks of Engineering

Parts of an Electrochemical Cell

A cell is a device that produces electrical energy. There are solar cells, photo cells, electrochemical cells and more, depending on what kind of energy is being converted into electrical energy.

Identify:

Which part of the electrochemical cell is…

1. The _____ is the negative electrode, where extra electrons gather.

2. The _____ is the positive electrode, where electrons leave from.

3. An _____ is where electrons can enter or exit a cell.

4. An _____ _____ produces electrical energy from chemical energy

5. A _____ cell has a liquid electrolyte.

6. A _____ cell has a solid electrolyte.

7. An _____ is responsible for removing electrons from the cathode and transferring them to the anode.

Analyze:

Why are the anode and cathode kept separate from each other?

Deduce:

Is there any contact between the anode and cathode at all? How do you know?

Labeling the Parts of an Electrochemical Cell

 A cell is a device that produces electrical energy. There are solar cells, photo cells, electrochemical cells and more, depending on what kind of energy is converted into electrical energy.

Label:

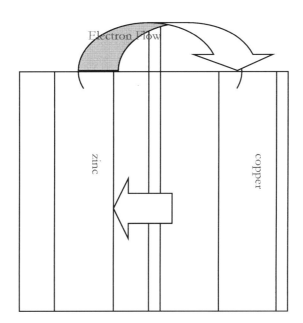

Label the above electrochemical cell with the following words:

A: Anode

B: Cathode

C: Electrolyte

D: Wire

Building Blocks of Engineering

Build a Battery Design and Discovery Lab

●	Design opportunity – **Build a battery**
●	Research- review what is inside an electrochemical cell
●	Brainstorm- write all your ideas below Consider what you will use for the:Anode, Cathode, Electrolyte, Battery covering
●	Sketch out your best idea below
●	Materials- list your materials below, and plan a procedure
●	Build it
●	Evaluate your design and how well it fulfilled the design opportunity
Additional Info	

Building Blocks of Engineering

Electrolyte Effectiveness

An electrolyte is a liquid connecting ions that can move freely within it carrying charge, which effectively allows for the conduction of electricity.

Design

Design an experiment that will allow you to test the effectiveness of different electrolytes.

Problem:

Hypothesis:

Materials:

Procedure:

Results:

Observations:

Conclusions:

Building Blocks of Engineering

Potato Battery Design and Discovery Lab

●	Design opportunity – **build a potato-powered circuit!**
●	Research- review what is inside an electrochemical cell
●	Brainstorm- write all your ideas below
●	Sketch out your best idea below
●	Materials- list your materials below, and plan a procedure
●	Build it
●	Evaluate your design and how well it suited the design opportunity
Additional Info	

Conductors and Insulators

 A wire is made with a conductor on the inside and an insulator on the outside.

Define

A conductor is a material that _____.

An insulator is a material that _____.

Sort:

Circle the conductors listed below.

aluminum	glass	ceramic	wool	gold
platinum	zinc	plastic	cotton	wood
twine	grass	leather	sand	foam
rubber	steel	water	paper	copper

Record:

As you test materials, list the conductors and insulators below:

Building Blocks of Engineering

STEM Middle School
Basic Electrical Engineering Lesson # 16

Circuit Necessities

 An open circuit will not have a complete path for electricity to flow from one end of the battery to the other.

Find Errors

Can you make a circuit with the following items? Why or why not?

Try each suggested circuit with the items from the kit and answer the questions below.

1. Two batteries, positive ends together

2. A battery, a light bulb, and one wire

3. Two wires and a light bulb

4. One wire and a battery

5. A wire, a switch, a battery, and a light bulb

6. Two wires, a switch, a battery, and a light bulb

7. Three wires, a switch, a battery, and a light bulb

Summarize:

For each component added to a circuit, how many wires need to be added?

Building Blocks of Engineering

Voltmeters

 Voltage, volts, and potential difference all refer to the same thing.

Evaluate:

1. Your voltmeter reads 1.2. What does this 1.2 stand for?

Predict:

Will voltage increase, decrease or stay the same?

1. adding another battery
2. adding a light bulb
3. adding a buzzer
4. adding a solar cell
5. using more wires
6. using a less efficient light bulb
7. using a more efficient light bulb
8. removing a battery
9. adding a fan
10. reducing current

Calculating Changes in Voltage Lab

Problem:

How is voltage affected by adding more light bulbs to a circuit?

Hypothesis:

Materials:

3 batteries

3 battery holders

3 light bulbs

3 light bulb holders

7 wires

1 switch

1 voltmeter

Procedure:

Build a circuit with no light bulbs and measure the voltage. Remember a voltmeter is never connected in series, only used in parallel.

Add one light bulb and measure the voltage.

Add a second light bulb and measure the voltage.

Add a third light bulb and measure the voltage.

Building Blocks of Engineering

Calculating Changes in Voltage Lab Continued

Results and Observations:

Circuit Components	Voltage	Brightness of bulb(s)
No light bulb		
1 light bulb		
2 light bulbs		
3 light bulbs		

Conclusion:

What did you learn about how adding more components to a circuit affects the voltage?

How does voltage affect the brightness of the bulb?

Extensions:

Is it true that adding **any** component to a circuit will decrease voltage? Support your answer with examples.

Ammeters

 Make sure your ammeter has an 'A' on it. If it has an 'mA 'on it, it is measuring milliamps, and you will have to divide your answer by one thousand before doing any Ohm's Law calculations.

Evaluate:

2. Your ammeter reads 32. What does this 32 stand for?

Predict:

Will current increase, decrease or stay the same?

1. adding another battery
2. adding a light bulb
3. adding a buzzer
4. adding a solar cell
5. using more wires
6. using a less efficient light bulb
7. using a more efficient light bulb
8. removing a battery
9. adding a fan
10. reducing current

Analyze:

What is the relationship between current and voltage?

STEM Middle School
Basic Electrical Engineering Lesson # 20

Calculating Changes in Current Lab

Problem:

How is current affected by adding more light bulbs to a circuit?

Hypothesis:

Materials:

3 batteries

3 battery holders

3 light bulbs

3 light bulb holders

6 wires

1 switch

1 ammeter

Procedure:

Build a circuit with no light bulbs and measure the current. Remember an ammeter can be connected in series.

Add one light bulb and measure the current.

Add a second light bulb and measure the current.

Add a third light bulb and measure the current.

Calculating Changes in Current Lab Continued

Results and Observations:

Circuit Components	Current	Brightness of bulb(s)
No light bulb		
1 light bulb		
2 light bulbs		
3 light bulbs		

Conclusion:

What did you learn about how adding more components to a circuit affects the current?

What is the relationship between current and voltage?

Extensions:

Is it true that adding **any** component to a circuit will decrease current? Support your answer with examples.

Building Blocks of Engineering

Ammeters

Resistance is analogous to friction. Friction is the opposition to movement, and resistance is the opposition to electricity.

Sort:

Circle the choice that would provide **less** resistance:

1. thick wire / thin wire

2. long wire / short wire

3. coiled wire/ straight wire

4. 35 degrees / 100 degrees

5. 44 gauge wire / 22 gauge wire

Analyze:

1. Why do you think NASA uses silver in communications components for their space ships?

2. Iphone chargers have notoriously short cords. Give a reason why.

3. Invent a situation where engineers would specifically use certain materials to create more resistance. Why would engineers **want** there to be resistance?

STEM Middle School

Basic Electrical Engineering Lesson # 22

Building Blocks of Engineering

Ohm's Law Problems

Remember: I = V/R

Restate:

Rewrite the equation to solve for V

Rewrite the equation to solve for R

Always write the formula you are using and show your work!

Calculate:

1. What is the current passing through a wire with a charge of 9 volts and a resistance of 3 ohms?

2. If a current of 40 Amps is passing through a wire with a resistance of 200 Ohms, what is the voltage?

3. 32 volts pass through a wire with an inherent resistivity of 8 ohms. Find the current.

4. If the resistance is greater than the voltage, will current ever be more than 1 amp?

5. Find the resistance of a circuit where 500 volts and a current of 50 amps are applied.

STEM Middle School
Basic Electrical Engineering Lesson # 23

Lab: Calculating Resistance from voltage and current

Problem:

How is resistance affected by adding more light bulbs to a circuit?

Hypothesis:

Materials:

3 batteries, 3 battery holders

3 light bulbs, 3 light bulb holders

6 wires

1 switch

1 ammeter

1 voltmeter

Procedure:

Build a circuit with no light bulbs and measure the current and voltage. Remember an ammeter can be connected in series and a voltmeter used in parallel.

Add one light bulb and measure the current and voltage

Add a second light bulb and measure the current and voltage

Add a third light bulb and measure the current ad voltage

Calculate the resistance using Ohm's law.

***You may need to convert your current from mA to A by dividing by 1000.

Lab: Calculating Resistance from voltage and current continued

Results and Observations:

Circuit Components	Current	Voltage	Resistance
No light bulb			
1 light bulb			
2 light bulbs			
3 light bulbs			

Conclusion:

1. What did you learn about how adding more components to a circuit affects the

   resistance? Why is this so?

2. What is the relationship between current and resistance?

3. What is the relationship between voltage and resistance?

STEM Middle School

Advanced Electrical Engineering Lesson # 1

Basic Circuit Building

A component is a part of a circuit, including wires, batteries and loads.

Identify

Which components are necessary to light a light bulb?

battery

resistor

switch

wire

diode

led

light bulb

light bulb holder

photocell

transistor

buzzer

fan

capacitor

Infer:

What could be some reasons that a light bulb is not lighting up?

Test:

Use your voltmeter to check all the batteries in your kit. Make sure they all have approximately the number of volts written on the side of the battery. Test each battery individually and give all the dead ones to your teacher.

STEM Middle School
Advanced Electrical Engineering Lesson # 2

Circuit Safety

Electricity is easily converted into heat, and too much heat can cause a very real, very dangerous fire.

Identify

Name 3 devices used to prevent electrical fires in household circuits.

Design

Draw the wires connecting the power plant to the lamp in your bedroom. Visualize how the electricity is brought to you.

Evaluate

Compare a glass fuse and a GFCI. How are they alike? How are they different?

Infer

What should you do in case of an electrical fire? Is throwing water on it the right thing to do? Why or why not?

Building Blocks of Engineering

STEM Middle School

Advanced Electrical Engineering Lesson # 3

Paths of Electricity

 Electricity will flow from the negative end of the battery to the positive end of the battery. If that flow is uninterrupted, the circuit is closed.

Evaluate:

Will the Lightbulb light up? Why or why not?

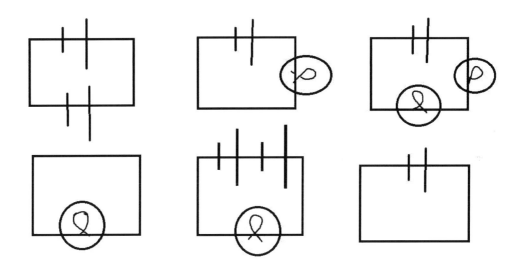

STEM Middle School
Advanced Electrical Engineering Lesson # 4

Schematic Diagrams

A component is a part of a circuit, including wires, batteries and loads.

Compose

What is a polarized component? List as many as you can below.

Create:

1. Draw the schematic diagram for a circuit with two batteries, a fan, a switch and wires below. Then, make it with your kit.

2. Draw the schematic diagram for a circuit with two batteries, a buzzer, a switch and wires below. Then, make it with your kit.

3. If you have time, go back to your diagrams and add a voltmeter and ammeter. Remember which is attached in series and which in parallel.

STEM Middle School
Advanced Electrical Engineering Lesson # 5

Light bulbs in Series Lab

 Series circuit have all the components connected in one large pathway.

Problem:

How will unscrewing one lightbulb affect the other 2 in the circuit?

Hypothesis:

Materials:

3 batteries

3 battery holders

Wires

3 lightbulbs

3 lightbulb holders

Procedure:

Build this circuit.

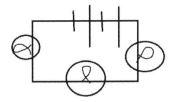

Light bulbs in Series Lab Continued

Results and Observations:

Conclusions:

Questions:

How many paths did the electricity have to flow through?

What would occur if you used a wire to bypass the broken/missing light bulb?

Building Blocks of Engineering

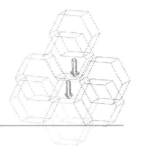

Building Blocks of Engineering

Light bulbs in Parallel Lab

 Parallel circuits have alternate paths for the electricity to flow.

Problem:

How will unscrewing one lightbulb affect the other 2 in the circuit?

Hypothesis:

Materials:

3 batteries

3 battery holders

Wires

3 lightbulbs

3 lightbulb holders

Procedure:

Light bulbs in Parallel Lab Continued

Results and observations:

Conclusions:

Questions:

How many paths did the electricity have to flow through?

Building Blocks of Engineering

Double Knife Switch Design and Discovery lab

●	Design opportunity – **Build a 2 volume buzzer using this double knife switch.**
●	Research- discuss with your teacher how the double knife switch works. Go over parallel and series circuits
●	Brainstorm- write all your ideas below
●	Sketch out your best idea below
●	Materials- list your materials below, and plan a procedure
●	Build it
●	Evaluate your design and how well it the design opportunity
Additional Info	

Building Blocks of Engineering

Double Knife Switch Design and Discovery lab

●	Design opportunity – **Build a REVERSE SWITCH** (when the switch is open the light bulb goes on and when the switch is closed the light bulb goes off)
●	Research- look at different reasons why a light bulb would not light in order to help you gather your thoughts. Also, review parallel circuitry.
●	Brainstorm- write all your ideas below
●	Sketch out your best idea below
●	Materials- list your materials below, and plan a procedure
●	Build it
●	Evaluate your design and how well it fulfilled the design opportunity
Additional Info	

Building Blocks of Engineering

Double Knife Switch Design and Discovery lab

●	Design opportunity – **Ever have that epic fight with your brother? You know, where you want the light on and he wants the light off? Well, imagine you have two light switches that control the same light bulb. Could you wire a circuit with only one light bulb and two switches, where every time either of the switches is lifted or closed, the light will turn on or off?**
●	Research- look at different reasons why a light bulb would not light, to help you gather your thoughts. Also, review parallel circuitry.
●	Brainstorm- write all your ideas below
●	Sketch out your best idea below
●	Materials- list your materials below, and plan a procedure
●	Build it
●	Evaluate your design and how well it fulfilled the design opportunity
Additional Info	

Building Blocks of Engineering

Switches

There are many types of switches, including regular single pole switches, double knife switches, and push-button switches.

Sort: Which switch would be most ideal for

1. an electric fencing match _____

2. a dual speed fan _____

3. a desk lamp_____

4. a game buzzer (to hit when you get the right answer) _____

5. an alarm system _____

6. a light dimmer _____

7. an "easy" button _____

8. a doorbell _____

9. a computer "on" button _____

10. the right-click button on a keyboard _____

STEM Middle School
Advanced Electrical Engineering Lesson # 11

Building a Pressure Plate

A push button switch works with the same essential mechanism as a pressure plate. Think about it.

Problem:

How can we build a pressure plate?

Hypothesis:

Materials:

3 batteries

3 battery holders

Wires

Aluminum foil

Sponge

Load

Procedure:

Use the design process to come up with an idea for a pressure plate. Sketch it below

Building Blocks of Engineering Curriculum All Rights Reserved

Building a Pressure Plate Continued

Results and observations:

How well did your pressure plate work? Any improvements you would make?

Conclusions:

Compare and contrast your pressure plate with the pushbutton switch from last lesson. How are they alike? How are they different?

Using Resistors in a Circuit

 Safety Data: Do NOT touch the resistor once it is part of a closed circuit. To touch the resistor, open the switch and let the resistor cool for 2-5 minutes.

Prelab Question:

State the law of conservation of energy.

Problem:

In what ways does a resistor affect a circuit?

Hypothesis:

Materials:

2 batteries, 2 battery holders

Wires

Light bulb

Light bulb holder

Switch

Resistor

Voltmeter

Ammeter

Thermometer (optional)

STEM Middle School
Advanced Electrical Engineering Lesson # 12 page 2

Using Resistors in a Circuit Continued

Procedure:

Build this circuit, then, add a resistor as shown in diagram 2. Note any changes in your results section.

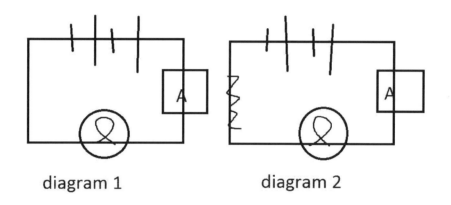

diagram 1   diagram 2

Results/Observations:

Properties	With no resistor	With a resistor

Conclusion:

What does a resistor do in a circuit?

Questions:

Engineers routinely place resistors in products such as laptops and cellphones. Why would they do this? Why is there a fan built into your laptop?

STEM Middle School
Advanced Electrical Engineering Lesson # 13

Resistor Color Bands

Color is placed on resistors since sometimes it is difficult to read the tiny numbers or letters on such a small component.

Calculate:

Band 1	Band 2	Band 3	Band 4	Upper limit	Lower limit
Tens digit	Ones digit	Multiplier	Tolerance	Value+ tolerance	Value - tolerance
Red	Red	Red	Silver		
yellow	Green	Blue	Gold		
Gray	orange	Orange	Silver		
green	Blue	Brown	Red		
blue	Brown	Black	Gold		
brown	Blue	Green	Gold		
purple	Brown	Gray	Gold		
red	Orange	Red	Red		
red	Green	brown	Silver		
yellow	gray	yellow	red		

Conclude:

Why don't engineers simply place numbers on the resistors?

What about the colors makes it more universal?

**Students will need to reference a color chart to calculate resistance based on color bands

STEM Middle School
Advanced Electrical Engineering Lesson #14

Adding Diodes to a Circuit

Safety Data: ALWAYS use a resistor in series with a diode, or your diode will burn out.

Prelab Question:

Draw the symbol for a diode. What symbol on your DVD player does it resemble? What two functions does that button have?

Problem:

In what ways does a diode affect a circuit?

Hypothesis:

Materials:

2 batteries, 2 battery holders

Wires

Light bulb, Light bulb holder

Switch

Diode, Resistor

Voltmeter

Ammeter

Adding Diodes to a Circuit Continued

Procedure:

1. Build the circuit in diagram 1, then add a diode and a resistor as shown in diagram 2. Note any changes in your results section.

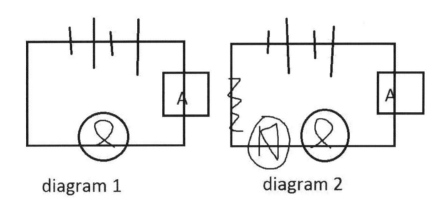

diagram 1   diagram 2

2. Now flip the diode around (change the polarization). Note any changes in your results section.

Results/Observations:

Properties	With no diode	With a diode	With the diode turned around

Conclusion:

What 2 functions can a diode have in a circuit?

Building Blocks of Engineering

Adding Diodes to a Circuit Continued

Adding Questions:

Name other polarized electrical components.

How do you think engineers make a diode have two different functions?

STEM Middle School
Advanced Electrical Engineering Lesson #15

Forward and Reverse Bias

 A diode is made of a semi-metal which is then "doped", meaning that other elements are added to it to alter its properties.

Distinguish:

Is this diode in forward bias or reverse bias?

1. the light bulb is on _____

2. the light bulb is off _____

3. there is high resistance across the diode _____

4. there is low resistance across the diode _____

5. current is flowing quickly throughout the whole circuit _____

6. current is slowing at one point in the circuit _____

7. in a "play" button _____

8. in a "pause" button _____

9. the negative end of the diode is connected to the negative end of the batteries

   _____

10. the negative end of the diode is connected to the positive end of the batteries

    _____

Create:

Draw a diagram of a diode in forward bias with a light bulb

Draw a diagram of a diode in reverse bias with a light bulb

Building Blocks of Engineering

Light Emitting Diodes

Safety Data: ALWAYS use a resistor in series with a diode or your diode will burn out.

Prelab Question:

Name polarized electrical components

Problem:

How do you connect an LED in a circuit?

Hypothesis:

Materials:

2 batteries

2 battery holders

Wires

LED

Switch

Voltmeter

Ammeter

Building Blocks of Engineering

Light Emitting Diodes Continued

 Safety Data: ALWAYS use a resistor in series with a diode or your diode will burn out.

Procedure:

Build this circuit, then add an LED as shown in diagram 2, and note any changes in your results section. Then flip the LED around (change polarization), note any changes in your results section.

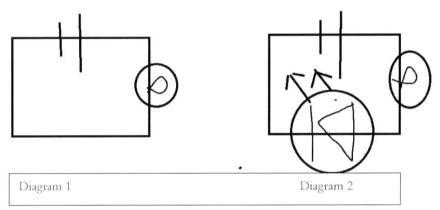

| Diagram 1 | Diagram 2 |

Results/Observations:

Properties	With no diode	With a diode	With the diode turned around

Conclusion:

How do you properly connect an LED in a circuit?

Building Blocks of Engineering

LED Design and Discovery Lab

●	Design opportunity – **Use LED's and button batteries to create a light-up invention! You may use cardboard, playdough, paper, string, glue, markers, tape, scissors, plates, bowls, and spoons. Happy inventing!**
●	Research- either look at different things that would be useful if they lit up, or research things kids would love to look at if they lit up.
●	Brainstorm- write all your ideas below
●	Sketch out your best idea below
●	Materials- list your materials below, and plan a procedure
●	Build it
●	Evaluate your design and how well it fulfilled the design opportunity
Additional Info	

Dance Dance Evolution Design and Discovery Lab

●	Design opportunity – **You have been hired by the "Dance DanceEvolution" company to create a new dance mat! Use your engineering prowess and electrical knowledge to design and build a dance mat that lights up in one of 4 different colors depending on what dance move you do!**
●	Research-
●	Brainstorm- write all your ideas below
●	Sketch out your best idea below
●	Materials- list your materials below, and plan a procedure
●	Build it
●	Evaluate your design and how well it fulfilled the design opportunity
Additional Info	

STEM Middle School
Advanced Electrical Engineering Lesson # 19

Electromagnetism Lab

Magnetism is a force that can push and pull. When two magnets push apart it is called repelling, and when two magnets pull together it is called attracting.

Problem:

How does wrapping a wire around a rod of metal and passing a current through it affect a circuit?

Hypothesis:

Materials:

3 batteries, 3 battery holders

Wires, Switch

Rod coiled with long wire

Procedure:

Build this circuit, and write down any changes you observe in both the rod and the circuit.

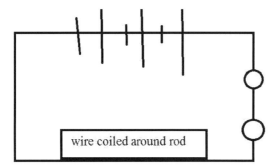

wire coiled around rod

Electromagnetism Lab Continued

 Magnetism is a force that can push and pull. When two magnets push apart it is called repelling, and when two magnets pull together it is called attracting.

Results/Observations:

Conclusion:

Building Blocks of Engineering

Electromagnetism and Coils Lab

Magnetism is a force that can push and pull. When two magnets push apart it is called repelling, and when two magnets pull together it is called attracting.

Problem:

How does the number of coils in an electromagnet affect its strength?

Hypothesis:

Materials:

3 batteries,3 battery holders

Wires

Switch

Rod coiled with long wire

Procedure:

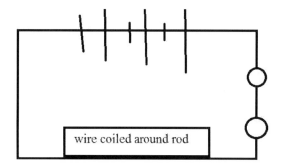

wire coiled around rod

Electromagnetism and Coils Lab Continued

Magnetism is a force that can push and pull. When two magnets push apart it is called repelling, and when two magnets pull together it is called attracting.

Procedure Continued:

1. Now coil the wire 40 times around the rod. Connect the circuit and try to lift paper clips. Record the number of clips you can lift.

2. Now coil the wire 60 times around the rod. Connect the circuit and try to lift paper clips. Record the number of clips you can lift.

3. Now coil the wire 80 times around the rod. Connect the circuit and try to lift paper clips. Record the number of clips you can lift.

Results/Observations:

	How many paper clips are lifted
80 coils	
60 coils	
40 coils	

Electromagnetism and Coils Lab Continued

Magnetism is a force that can push and pull. When two magnets push apart it is called repelling, and when two magnets pull together it is called attracting.

Conclusion:

How does the amount of coils in an electromagnet affect the strength of that electromagnet? How do you know?

Extension:

How do you think the thickness of a rod would affect the strength of an electromagnet? What experiment could you do to determine the answer?

How else would you test the strength of an electromagnet?

Building Blocks of Engineering

Building Blocks of Engineering

Electromagnetism Lab

 A component is a part of a circuit. In each mystery box you will find a different component, identified only by the effect it has on your control circuit.

Solve:

1. Mystery Box #____

   Contains a _____.

   I know this because _____.

   This is my diagram of the circuit:

2. Mystery Box #____

   Contains a _____.

   I know this because _____.

   This is my diagram of the circuit:

3. Mystery Box #____

   Contains a _____.

   I know this because _____.

   This is my diagram of the circuit:

Building Blocks of Engineering

Arcade Game Design

	Design opportunity – **Make your own arcade game!**
	Research- look at handheld arcade games and review the components you have learned so far.
	Brainstorm- write all your ideas below
	Sketch out your best idea below
	Materials- list your materials below, and plan a procedure
	Build it
	Evaluate your design and how well it fulfilled the design opportunity
Additional Info	

Building Blocks of Engineering

STEM Middle School
Electrical and Computer Systems Lesson # 1

Systems:

 A system is a set of parts and procedures that interact to accomplish a goal

Define the Relationship

1. system
2. input
3. output
4. feedback
5. open-loop system
6. closed-loop system
7. automated system
8. manual system

9. indicator
10. sensor

a. object that detects a change in a system
b. system controlled by human interaction
c. device reporting info on state or condition
d. a system that constantly requires new input
e. output that is put back into the system as new input
f. a system with little or no human interaction
g. information given or put into a system
h. set of parts and procedures that work together to accomplish a task or set of tasks
i. information produced by a system
j. a system that uses output as new input

Identify the system

(choose closed or open loop and manual or automatic)

1. a breathalyzer test that beeps if there is a high alcohol content in your breath
2. a thermostat that adjusts without you doing anything
3. a hoverboard that adjusts to your position
4. a videogame that requires you to press buttons to achieve certain actions.

STEM Middle School
Electrical and Computer Systems Lesson # 2

Open-loop system Design and Discovery lab

●	Design opportunity – **Build a maze out of silver foil that is narrow enough to make it hard for the player to move a metal rod through it without touching the sides. Make sure that if the player touches the rod to the silver foil maze, an indicator will go on.**
●	Research- aerodynamic shapes, ways to reduce drag and increase thrust
●	Brainstorm- write all your ideas below
●	Sketch out your best idea below
●	Materials- list your materials below, and plan a procedure
●	Build it
●	Evaluate your design and how well it fulfilled the design opportunity
Additional Info	

Vibration Sensor Design and Discovery Lab

●	Design opportunity – **Build a <u>vibration sensor</u>- a device that can detect strong motion and cause an indicator to go on when the motion is detected**
●	Research- aerodynamic shapes, ways to reduce drag and increase thrust
●	Brainstorm- write all your ideas below
●	Sketch out your best idea below
●	Materials- list your materials below, and plan a procedure
●	Build it
●	Evaluate your design and how well it fulfilled the design opportunity
Additional Info	

Building Blocks of Engineering (vertical left margin)

Safe Alarm Design and Discovery Lab

●	Design opportunity – **Build an alarm that will turn on when the safe door is opened and shut off when the safe door is closed.**
●	Research- aerodynamic shapes, ways to reduce drag and increase thrust
●	Brainstorm- write all your ideas below
●	Sketch out your best idea below
●	Materials- list your materials below, and plan a procedure
●	Build it
●	Evaluate your design and how well it fulfilled the design opportunity
Additional Info	

(vertical text, left margin) Building Blocks of Engineering

How to Use PowerPoint

The Power of the PPT

BB Curriculums
Unit 4
Lesson 5

Save to a flashdrive

- File
- Save As
- My Computer
- Locate your flashdrive (if you have no idea what it is called, unplug it and stick it back in and watch for the name that disappears and then pops back up!)
- Save this now in a folder called English_PPT

Building Blocks of Engineering

How to Use PowerPoint

The slide bar
(it's not a dance)

- Left side
- Move between slides easily
- Select a slide and move it ahead or behind of another one.

Set a background color

- Format
- Background
- Either apply it to all slides or set it separately for each slide
- Make the background a pale blue

Building Blocks of Engineering

How to Use PowerPoint

Choose a layout you love

- Format
- Slide layout
- Choose the layout with a title and bullet points. Each time you press enter, a new bullet point begins.

Add an extra slide

- Cntrl-m
- Try it!

Building Blocks of Engineering

How to Use PowerPoint

Format Text

- You can use a text box, but they are often hard to read
- Insert-Picture-Word Art for **fancy cool lettering**
- Select text and Format-Font for options to change font, size, color, and other aspects of text.
- Make this text green, size 34 , helvetica

Insert a Picture

- Insert (no kidding)
- Picture (shocker, shocker)
- If you are using your own picture-select "from file"
- If you would like clip art stored in the software select "clipart", and a clipart bar will appear on the right side of your screen
- Add a clipart picture here!

How to Use PowerPoint

ANIMATION!!!

- This is where PowerPoint Presentations get super cool!
- To animate an object:
  - Highlight or select the text or object
  - Slideshow
  - Custom Animation
  - Add Effect

Types of Animation

- Entrance
- Emphasis
- Exit
- Motion Path
- Add a cool animation here!

Building Blocks of Engineering Curriculum All Rights Reserved

STEM Middle School
Electrical and Computer Systems Lesson # 6

Make your own PowerPoint Presentation on an Automatic System

 An automatic system works with little or no human interaction.

Sample Outline:

Title Slide-

Name, Class, Title, Subtitle

Introduction Slides-

What is an automatic system?

Your system slides – Research an automatic system

1. overview
2. how does the system turn on (is there any human interaction)?
3. what is the input?
4. what feedback is produced?
5. which sensor process the feedback?
6. other

Conclusion-

1. any problems with the system

2. possible improvements

3. similar systems

STEM Middle School
Electrical and Computer Systems Lesson # 7

Basic Computer Terms

 A system is a set of parts and procedures that interact to accomplish a goal.

Determine the Correct Term

Choose the correct answer from the choices below:

1. _____ - the physical parts of a computer system that you can touch, includes the monitor, keyboard, all internal physical components.
2. _____ - the programs and procedures that a computer carries out
3. _____ -also known as a circuit board, it facilitates communication between different hardware pieces by providing electrical connectivity.
4. _____ - a central processing unit receives all the instructions generated by the software and executes the correct actions in response to the commands.
5. _____ - another name for the CPU, it is likewise responsible for manipulating data according to software instructions
6. _____ - also called storage, where data is stored either in electronic (digital) or optical form until the computer needs to access it.
7. _____ - random access memory- refers to an actual hardware component which stores information on your computer.
8. _____ - a unit used for measuring amounts of information stored by a computer
9. _____ - the smallest piece of an image as stored by a computer or digital device
10. _____ -the number of pixels in an image
11. _____ - it is basically a program that controls a device. It is a piece of software that allows your computer to interact with different kinds of hardware, for example an iphone or a memory card.
12. _____ - a collection of instructions or commands for a system
13. _____ – writing a computer code

Word Box

CPU	-Core Processor	-Pixel	-Computer Code	-Memory	-Programmer	-RAM
-Software	-Motherboard	-Driver	-Byte	-Hardware	-Resolution	

Building Blocks of Engineering

Image Compression

Represent a picture through letters and numbers!

For example: look at the picture below. It could be represented as follows:

5

1,2,2

1,2,2

1,2,2

0,4,1

The computer will alternate coloring boxes white or black. It will always begin with white, so if the first box is black, precede the code with a zero. Each number represents how many boxes should be filled in.

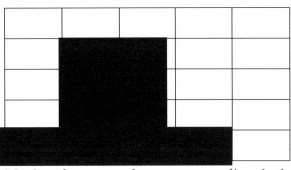

Notice the empty boxes are not listed, the default is empty.

Draw a picture in the grid above. Fill in boxes either entirely, half-way or not at all.

Image Compression

 A system is a set of parts and procedures that interact to accomplish a goal.

Create your own system:

Create your own system for letting someone know how to fill in a grid so that they draw the picture you had in mind. Then give your friend the instructions and a piece of graph paper and get to work!

Record your instructions below:

STEM Middle School

Electrical and Computer Systems Lesson # 10 and 11

Stix Fix Programming Basics

 An automated system will follow your commands in the sequence they are given.

Learn a New Language!

You will now be creating a programming language which we will call "Stix".

Using the language of Stix, you will write out instructions for a certain construction using popsicle sticks with ridges.

You and your partners will take turns.

When you are the coder, you will be in charge of writing the correct "code", set of instructions, that your partner will follow in order to achieve the correct construction.

Your partner will then be playing the role of "robot", where he or she will follow the commands you have written exactly.

Often, you will notice a flaw in your instructions based on the construction flaw you see when the "robot" carries out the instructions. No problem! Go back and "debug" the code and let the robot try and "run the program" again!

The class will be working with a certain set of commands (such as "pick up stick" or "put stick down"), but if you feel another command would be useful, feel free to ask the teacher to add it to the Stix language.

Stix Commands for Today's Lesson:

It is important to note that when you pick up a stick, it is always assumed to be in this position.

Command	Symbol
1. Pick up a stick	
2. Put down a stick	
3. Turn a stick 90 degrees up to the left	
4. Turn a stick 90 degrees up to the right	
5. turn a stick 90 degrees down to the left	
6. turn a stick 90 degrees down to the right	
7. move a stick over #_ ridges	Over ( )

Stix Challenge 1:

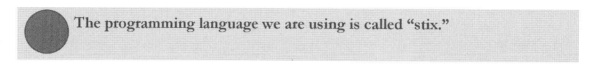

The programming language we are using is called "stix."

Build an interlocking square:

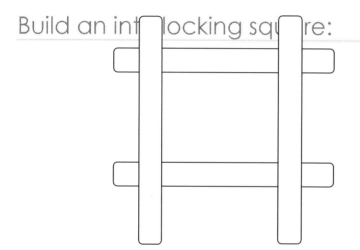

Write your code below!

Connect:

Look at how your robot is performing the tasks exactly as you have coded them. How is this an example of automation?

Reverse engineering

Design

Come up with the sequence of code necessary to get your robot to build this structure: (a beam bridge).

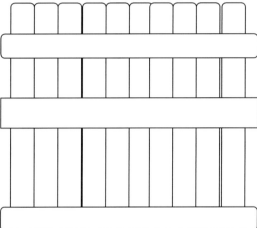

Write Your Code Below:

STEM Middle School
Electrical and Computer Systems Lesson # 13

Computing and Sequencing

 A sequence is a set of objects events or commands that flow in a linear pattern.

Critical Thinking

In what order will the robot follow your commands?

Does that order make a difference?

Work Backwards:

Go back to last weeks' code and have the robot follow it backwards as best as he can.

Draw Conclusions:

What difference did it make that the sequence was disturbed?

STEM Middle School
Electrical and Computer Systems Lesson # 14

Repetition and algorithms

 Coding is meant to be efficient and concise.

Create

Build a pyramid that is three layers of boxes across.

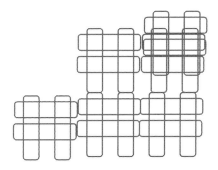

Assess:

Do you notice any repetitions in the code? Write them below.

Respond:

What is a more efficient way to write the repetitive code?

STEM Middle School
Electrical and Computer Systems Lesson # 15

An Hour of Code

 Coding helps develop your problem-solving and critical thinking skills.

Build an Understanding:

How many commands in total are there in the lightbot game?

How many different structures was the robot able to climb?

What important coding strategy do you need to use in order to condense code?

Can you use an algorithm inside another algorithm?

STEM Middle School
Electrical and Computer Systems Lesson # 16

User Input

User input makes it possible for websites to customize the pages you are viewing.

Enrich:

Add the following commands to your Stix command Bank:

Ask () – use this command to ask a user/player a question by writing it on a stick

Display ()-use this command if you would to write information on a stick

Answer- use this to refer to the answer given by the user

Create:

Create a code that instructs you to build a house and then ask the user what their name is. When the user tells you their name, instruct the robot to write it on a stick and place it on top of the house.

Building Blocks of Engineering

Conditional Programming

A computer is able to check if a condition is true or false before carrying out another command.

Enrich:

Add the following commands to your Stix command Bank:

If ( ) then ( )- use this command to perform an action only if a certain condition is met.

Create:

Ask your robot to build a large fence with a door.

Have your robot display "you must knock before entering" on the door.

Now, program your robot that:

> *If (user knocks) then (open the fence door)*

STEM Middle School
Lesson # 18

More conditional Programming

A computer is able to check if a condition is true or false before carrying out another command.

Enrich:

Add the following commands to your Stix command Bank:

Display ( ) – use this command if you would to write information on a stick

If ( ) then ( ) otherwise ()- use this command to perform an action only if a certain condition is met, and if the condition is not met, to perform a second action.

Create:

Ask your robot to make two boxes, and display "correct" on one box and "incorrect" on the other box.

Have your robot "ask" the user to guess a number between 1 and 10.

Now, program your robot to place a stick in the "correct" box if the number is 4, and in the incorrect box if the number is not equal to 4.

STEM Middle School
Electrical and Computer Systems Lesson # 19

Parallel Execution

 A computer is able to work efficiently by performing many tasks per second.

True or False:

1. Parallel execution requires more effort on the part of the processor(s) _____
2. Parallel execution is less efficient _____
3. Two distinct activities cannot be carried out at the same time by one processor __
4. Parallel execution saves time _____
5. A computer can carry out many many processes at the same time _____

Create:

Write the code for building a staircase out of Popsicle sticks below.

Experiment:

1. Now build it yourself and record the time it takes you to complete it.
2. Ask your partner to help you and record the time it takes both of you to do it together.

Who is building	Time
Just you	
You and a partner	

Conclude

Draw a conclusion about the effects of parallel execution on the time it takes to execute a task.

STEM Middle School
Electrical and Computer Systems Lesson # 20

Programming with Variables

 A computer is able to check if a condition is true or false before carrying out another command.

Enrich:

Add the following commands to your Stix command Bank:

Create variable ( )

Set variable ( ) = to ( )

A variable is a named quantity.

Create:

1. Build a basket out of sticks.

2. Write the code below.

3. Create the variable called "score"

4. Instruct your robot to display the variable "score"

5. Set score equal to the number of balls thrown in the basket.

6. Now create some basketballs out of crumpled paper and invite some friends to shoot some hoops.

7. Allow each friend to try and shoot 5 baskets.

7. Your robot should be able to display the score as a number of 0-5.

Building Blocks of Engineering

STEM Middle School
Electrical and Computer Systems Lesson # 21

Programming Vocabulary

 Coding is a language everyone can understand.

Matching:

1. code	a. named quantity
2. programmer	b. information provided by a user
3. automation	c. procedures that will be executed only if certain information is true
4. sequence	d. instructions for a computer program
5. parallel execution	e. a code that instructs a program to carry out a sequence of commands
6. synchronization	f. two processes being carried out at once
7. user input	g. a repetition that repeats forever
8. variables	h. a sentence that computes using a variable
9. repetitions (algorithms)	i. someone who writes code
10. loops	j. when events are timed based on other events
11. conditional programs	k. the computer processing without human action
12. expressions	l. a specific order of computer code

Process:

Write 3 true or false statements regarding any of the programming vocabulary listed above.

Building Blocks of Engineering

Stix Game Design

●	Design opportunity – **Create the code for a program that will instruct the robot to build or play a game. You may use a new game of your design or have the robot build a pre-existing one.**
●	Research- aerodynamic shapes, ways to reduce drag and increase thrust
●	Brainstorm- write all your ideas below
●	Sketch out your best idea below
●	Materials- list your materials below, and plan a procedure
●	Build it
●	Evaluate your design and how well it fulfilled the design opportunity
Additional Info	

Building Blocks of Engineering

STEM Middle School
Mechanical Engineering I Lesson # 1

Newton's First Law of Motion

 Task: Place an index card over the top of a plastic cup. Place a coin or other small object on top of the index card. Now, pull the index card out very quickly.

1. What happens to the coin?

2. Why does this happen?

Inertia

1. A car is going 20 miles an hour when it suddenly stops short. The passenger is wearing a seatbelt. What would happen without the seatbelt? Why?

2. What would occur if you rolled a marble down a hallway? When would it stop?

3. What if you would shoot the marble out of a gun in outer space? When would that marble stop?

4. What is the unbalanced force that makes rain fall downwards?

Extensions

 Come up with your own example of inertia in daily life:

STEM Middle School
Mechanical Engineering I Lesson # 2

Newton's Second Law of Motion

Remember: Force = Mass*Acceleration

1. Solve the equation for Mass:
2. Solve the equation for Acceleration:

Calculate

1. Is it harder to push a full shopping cart or an easier one? How do you know?

2. If both a baseball and a much larger dodgeball are each hit with 25 N of force, which will accelerate more? Prove this with math!

3. How much force will it take to cause a 2000 g car to accelerate from rest to .5 m/s²?

Integrate

Roll a golf ball down an inclined plane and measure how quickly it reaches a certain point on the floor. Now, try it with a much lighter ping –pong ball. Which goes faster? Why is this true?

Building Blocks of Engineering

Newton's Third Law of Motion

Remember: Every action has an equal and opposite reaction.

1. What does a spaceship need in order to lift off?

2. If there is no air to maneuver in space, how can a spaceship move?

Experiment

- Blow up a balloon, but do not tie it. Instead, clip it with a clothespin.
- Decorate and label it.
- Tape a straight (unbent) plastic straw to the top of the balloon.
- Thread a string through the straw.
- Have two people hold the ends of the string and move the balloon all the way to one side.
- Open the paper clip.
- Record what happens below:

Conclude

Explain the movement of the balloon:

Building Blocks of Engineering

Design and Discover Lab

●	Design opportunity — **Build a Newtonian powered cart using a small car or block of wood, string, rubber bands, nails, tape and scissors.**
●	Research- Newton's laws
●	Brainstorm- write all your ideas below
●	Sketch out your best idea below
●	Materials- list your materials below, and plan a procedure
●	Build it
●	Evaluate your design and how well it fulfilled the design opportunity
Additional Info	

Gravity

 Remember: Gravity is a universal force based on mass.

Calculate

1. Tony proposed the following: If I take a paper and add nothing to it and do not cut it in any way, I can shape it so that it will fall more quickly than any other paper. What do you think Tony did with his paper?

2. If a ball weighing 10 kg begins to fall off a cliff, after 3 seconds how fast will it be going?

3. How do you know that the moon is affected by earth's gravity?

4. What two forces will cause an object to begin orbiting another object?

5. If you have two spheres that are dropped at exactly the same time and both spheres are exactly the same size and shape, but one weighs twenty pounds and the other ways one hundred pounds, which will hit the ground first?

Differentiate

 Imagine you lived on the moon, would you weight more or less? Why?

Kinetic and Potential Energy

 Remember: There are seven types of energy: electrical, chemical, nuclear, sound, electromagnetic, thermal, and mechanical.

Write kinetic or potential

1. A car drives down a hill _____
2. A ballerina is poised to pirouette _____
3. a fire burns _____
4. a match sits in the box _____
5. a spring is coiled _____
6. a lever is pulled back _____
7. a man stands at the edge of a cliff _____
8. hydrogen and oxygen bond to form water _____
9. a toy is wound up _____
10. a screen lights up _____

Experiment

Use a piece of string and some weights to make a pendulum. Study the movement of the pendulum and answer the following questions.

1. When does the pendulum have the most potential energy?
2. When does the pendulum have the least potential energy?
3. When does the pendulum have the most kinetic energy?
4. When does the pendulum have the least kinetic energy?

Building Blocks of Engineering

Design and Discover Lab

●	Design opportunity – **Build a roller coaster using flexible tubing and marbles. You may use other materials if you wish. You will also need to explain where the marble has kinetic energy and where it has potential energy.**
●	Research- Review Newton's 3 laws, especially his Third Law
●	Brainstorm- write all your ideas below
●	Sketch out your best idea below
●	Materials- list your materials below, and plan a procedure
●	Build it
●	Evaluate your design and how well it fulfilled the design opportunity
Additional Info	

Building Blocks of Engineering

Gravity

Remember: Work = Force x Distance.

Compute

1. You pull your trashcan through the snow for a distance of 250 m with a force of 20 N. How much work did you do?

2. The snow gets too deep, so you decide to carry your trash bag the rest of the way to the curb. How much work did you do when you lifted the trash bag which weighs 80 N, 2 meters into the air?

3. If you lift a 400 N backpack while doing 1200 J of work, how high did you lift the backpack?

4. How much does a box weigh if you did 500 J of work to carry it just 25 meters?

5. A bird sits on a rhino's nose. The rhino goes 100 miles in search of food, not even knowing the bird was there. Did the rhino do any work?

6. A boxer decides to test his strength by pushing against a pillar. He stands still and pushes with 300 N of force. How much work did he do?

Efficiency

Remember –

The efficiency of a machine can be worked out using the equation:

$$\text{Efficiency} = \frac{\text{useful energy output}}{\text{total energy input}}$$

True or False:

1. You can end up with more energy than you started with
2. Energy must always come from a source; it cannot be created
3. Simple machines use several types of energy
4. A perfect machine has a greater output than input
5. Real machines always lose some energy to friction/heat

Calculate

1 An engine has an energy input of 100 J and gives out 37 J of useful kinetic energy. What is the efficiency of the engine?

_____

_____

_____

2 A hair dryer has an energy input of 1600 J and gives out 1200 J of useful energy. What is the efficiency of the hair dryer?

_____

_____

_____

Building Blocks of Engineering (sidebar)

Efficiency Continued

3   Compare these two machines to decide which is more efficient.

     i   Machine A uses 1000 J of energy to do 400 J of useful work.

     ii   Machine B uses 500 J of energy to do 200 J of useful work.

4   Compare these two cranes to decide which is the more efficient.

     i   Crane A uses 800 J of energy to do 350 J of useful work.

     ii   Crane B uses 750 J of energy to do 300 J of useful work.

Synthesize

 Explain what conditions a perfect ideal machine would operate under.

STEM Middle School
Mechanical Engineering I Lesson # 10

Simple Machines

Remember: There are 6 simple machines; lever, wedge, screw, pulley, inclined plane and wheel and axle.

Which simple machine is this?

1. knife
2. seesaw
3. fan blade
4. doorknob
5. light switch
6. windowshade strings
7. soda bottle cap
8. wheelchair ramp
9. elbow joint
10. hinge
11. slanted bottom of a bath tub
12. roller blades
13. push pins
14. baseball bat
15. scissors

Detect

Look around the room. What other simple machines can you find?

Mechanical Advantage

Remember: What formulas are you working with?
   MA =
   Effort =
   Maximum Resistance =

Match the terms

1.	Resistance	a. the amount of times a machine multiplies a force
2.	Load	b. the amount of force you input
3.	Effort Force	c. what you are trying to lift
4.	Mechanical Advantage	d. another name for resistance
5.	Efficiency	e. useful work minus what is lost to heat and sound
6.	downward arrow	f. symbol used in diagrams for resistance
7.	box	g. used to measure force
8.	Newton	h. symbol used in diagram for effort

Solve

1. If the MA is 15, how much effort does John need to use to lift a 90 N gorilla?

2. If Debbie can lift a 500 N chocolate bar with only 5 N of force, what is the MA of whatever machine she is using?

3. What is the maximum resistance Alan can lift using his machine, which has an MA of 10 and an effort of 44 N?

STEM Middle School
Mechanical Engineering I Lesson # 12

Inclined Plane Discovery Lab

In today's lesson you will be using a spring scale to drag an object over several different inclined planes. You will measure the force required to drag that object up the inclined plane, and compare your results in order to determine how the height to length ratio (also called slope or incline or steepness) affects the force required to move an object.

Procedure

1. Attach a 50 g weight to the end of your spring scale.
2. Use a ruler to measure 10 cm.
3. Hold one end of the textbook to the 10 cm mark, and drag the weight up the textbook. Record the force it used below.
4. Repeat for 20 cm, 30 cm, 40 cm, and 50 cm.

Results

Height of inclined plane	Force required
10 cm	
20 cm	
30 cm	
40 cm	
50 cm	

Conclude

 Explain the force used:

Building Blocks of Engineering

Mechanical Advantage

Remember: What formulas are you working with?
MA of an Inclined Plane

MA= length of incline/ height

Compute

Complete the chart below:

MA	Length of incline (m)	Height (m)
2	30	15
3	30	
8	80	
	14	2
	72	12

Solve

If an inclined plane has a length of incline that is 60, and a height that is 5, how much effort will it take to move a resistance of....

Effort force required	Resistance
	36
	84
	120

Building Blocks of Engineering

Mechanical Advantage Continued

Complete this chart:

Length of incline	Height	M.A.	# of times force is multiplied	Resistance	Effort Force
40	10	4	4	88	22
45	9			500	
75	25				50
22	11			300	
7	7			30	

Building Blocks of Engineering

STEM Middle School
Mechanical Engineering I Lesson # 14

Mechanical Advantage Word Problems

Remember: What formulas are you working with?
MA of an Inclined Plane

MA= length of incline/ height

1. Which inclined plane will allow you to put in less effort, the one with a height of 5 and a length of incline of 60 or the one with a height of 12, and a length of incline of 60?

2. Mary is pushing a 66 N carton up an incline length of 8 m.. The height of the ramp is only 2 m. How much force does Mary need to exert?

3. Using a force of 55 N and an incline plane with a height of 10 meters and a length of incline of 100 meters, what is the maximum resistance Mary can lift?

Mechanical Advantage Word Problems Continued

Remember: What formulas are you working with?
MA of an Inclined Plane

MA= length of incline/ height

Using the inclined plane below, answer the following questions:

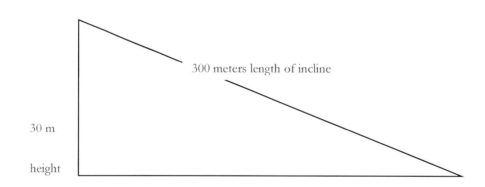

300 meters length of incline

30 m

height

1. How much effort will Noah have to use to push a 45 g box up the ramp?

2. Draw and label an inclined plane with more MA

3. Draw and label an inclined plane with less MA

Building Blocks of Engineering

Mechanical Advantage Word Problems Continued

Remember: There are 3 classes of levers:

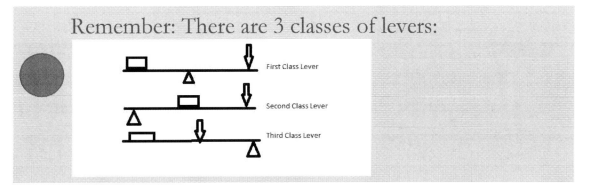

Classify the kind of lever listed below

1. broom
2. elbow
3. pair of scissors
4. fishing rod
5. baseball bat
6. door hinge
7. can opener
8. crow bar
9. seesaw
10. wheelbarrow
11. tweezers
12. nutcracker
13. hinge
14. stapler
15. rake
16. oars on a rowboat
17. tongs
18. catapult
19. hammer pulling out a nail
20. tennis racket

Building Blocks of Engineering

Design and Discover Lab

●	Design opportunity –**Build a catapult!**
●	Research- review first, second, and third class levers.
●	Brainstorm- write all your ideas below
●	Sketch out your best idea below
●	Materials- list your materials below, and plan a procedure
●	Build it
●	Evaluate your design and how well it fulfilled the design opportunity
Additional Info	

Lever Discovery Lab

Define "effort arm" or "effort distance":

Define "resistance arm" or "resistance distance":

Problem:

How does the length of the lever (effort arm) affect the amount of force required to lift a load?

Hypothesis:

Materials:

Ruler (30 cm)
Pencil
Several pennies
One metal washer or small weight

Procedure:

1. Place the pencil directly beneath the middle of the ruler, creating equal effort and resistance arms.
2. Place the weight (load) on one end on the lever
3. Add pennies to the other end of the ruler until the weight is lifted.
4. Record your results below
5. Slide the pencil down 5 cm in order to make the effort arm 5 cm longer and the resistance arm 5 cm shorter.
6. Add pennies to the other end of the ruler until the weight is lifted.
7. Record your results below
8. Slide the pencil down another 5 cm and repeat steps 6 and 7
9. Slide the pencil down another 5 cm and repeat steps 6 and 7

Lever Discovery Lab Continued

Results:

Length of effort arm	Pennies required
15 cm	
20 cm	
25 cm	
30 cm	

Conclusions:

How does the length of the effort arm affect the amount of force required to lift the load?

MA of LEVERS

Remember: MA= Effort arm/ Resistance arm

Effort arm x Effort = Resistance arm x Resistance

MA= Resistance/Effort

Complete the Chart Below:

MA	Effort Arm cm	Resistance Arm cm
5	100	20
3	90	
	50	5
	36	12
4		20

Using this lever....

1. Using a lever with an effort arm of 60 cm and a resistance arm of 20 cm, how much effort does Moshe need to exert in order to lift 75 kg?

2. Using a lever with an effort arm of 300 cm and a resistance arm of 200 cm, which side should Ellen put her effort force on? Why?

3. Which way will this lever rotate? The left side has a weight of 45 kg and a length of 4 m, while the right side has a weight of 80 kg and a length of 2m. Show your calculations.

MA of LEVERS (Effort and Resistance Arm)

Remember: **MA= Effort arm/ Resistance arm**

Effort arm x Effort = Resistance arm x Resistance

MA= Resistance/Effort

Complete the Chart Below:

MA	Effort Arm cm	Resistance Arm cm
20		20
	75	5
	320	5
4		12
2	200	

Infer Using The Following Lever:

1. Using a lever with an effort arm of 25 cm and a resistance arm of 10 cm, how much effort does Sean need to exert in order to lift 75 N?

2. Using a lever with one arm with a length of 300 cm and another arm with a length of 600 cm, which side should Deborah put her effort force on? Why?

MA of a Single Fixed Pulley Discovery Lab

 Remember how to use a spring scale?

Problem:

How does using a single fixed pulley to lift a load affect the amount of force required to lift the load?

Hypothesis:

Materials:

Somewhere to hook the pulley
Spring scale
Some 50 g weights
A single pulley
Some string

Procedure:

1. Lift the weights with the spring scale and record how many Newtons of force it required to lift each set to a given height (it is best to lift the height you are hanging your pulley at)
2. Record your answers in the chart below
3. Ask your teacher to show you how string the pulley
4. Now attach the weights to one end of the pulley
5. Attach the spring scale to the other end of the pulley, and use the spring scale to pull the weights up to the given height
6. Record your results below
7. Repeat for each set of weights as given in the chart below

MA of a Single Fixed Pulley Discovery Lab Continued

Results:

Weight	Force required to lift without a pulley	Force required to lift it with a pulley
50 kg		
100 kg		
150 kg		
200 kg		
250 kg		

Conclusions:

Questions:

Are there any (other) advantages to using a single pulley?

Building Blocks of Engineering

MA of Pulleys

Type of pulley	MA
Single Fixed	Always 1
Single Movable	Always 2
Block and tackle	Count the strings (work divided among strings evenly)

Analyze

1. How much effort is required to lift a 25 N brick using a single fixed pulley?

2. How much effort is required to lift a 25 N brick using a single movable pulley?

3. How much effort is required to lift a 25 N brick using a 5 stringed block and tackle?

4. What is the main advantage of using a single fixed pulley?

5. Can a single movable pulley change the direction of the force?

6. Can a block and tackle multiply a force?

7. Can a block and tackle change the direction of a force?

8. How much resistance can Ben lift if he exerts 22 N of force pulling on his 3-stringed pulley?

9. How much rope does Ronit need to pull in order to lift her 2 stringed pulley 4 m off the ground?

Efficiency and Pulleys

REMEMBER: Efficiency= energy out divided by energy in.

Complete the Table Below:

Energy Output	Energy Input	Efficiency
1. 285 J	400 J	
2.	100 J	90%
3. 260 J		55%
4. 840 J	850 J	
5. 20 J		20%

REMEMBER to use your mechanical advantage formulas! Look them up!

Calculate:

1. Sarah ties a 50 N bucket to one end of a pulley with 5 strings. How much force will she need to use to lift the bucket up to the tree house, assuming her pulley is 90% efficient? [use actual force in = (load / MA) / Efficiency]

2. Calculate the ACTUAL mechanical advantage from problem one. [Use AMA= load/actual force in]

3. Considering Sarah does 88 N*m of work while exerting the force from problem one, how many meters of rope did she need to pull? [Use W=FxD]

Efficiency and Pulleys Continued

 A pulley is like any other machine. It can multiply force, but it also multiplies distance.

Infer:

# of strings on Pulley	Resistance (load)	Resistance on each string	MA your force is multiplied by a factor of…	Distance multiplied by a factor of…
Example: 5	150 N	30 N	5	5
1.		4 N	12	
2. 11	330 N			
3.		15 N	2	
4. 6		5 N		
5.	18 N			3

STEM Middle School
Mechanical Engineering I Lesson # 23

Design and Discover Lab

	Design opportunity – **Build a machine that can lift 1000 g using a force of only 5 Newtons!**
	Research- levers, inclined planes and wheels and axles.
	Brainstorm- write all your ideas below
	Sketch out your best idea below
	Materials- list your materials below, and plan a procedure
	Build it
	Evaluate your design and how well it fulfilled the design opportunity
Additional Info	

What is a Screw:

Task: Cut out the following triangles and wrap them around a pencil starting with the edge that has the arrow. What do you see?

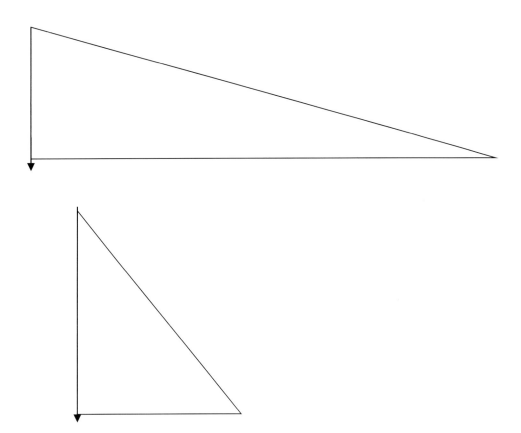

Compare and Contrast:

Compare and contrast the two different screws you have made. What is similar about them? What is different?

Screw Discovery Lab

Remember: A screw is an inclined plane wrapped around a rod.

Problem:

What properties of screws make them easier or harder to screw into a piece of wood?

Hypothesis:

Materials:

Block of wood
Several different screws
A nail
A hammer
A screwdriver

SafetyData:

Be careful of fingers when using hammer and handling nails and screws.
Safety goggles can be worn as a precaution

Procedure:

10. Use the hammer to hammer a nail into the wooden block.
11. Record how much effort it took from a scale of 1-10 (1 being the least and ten being the most)
12. now choose the screw with the fewest grooves
13. Use the screwdriver to screw that screw into the board. Record how many complete turns it took and rate how much force it took.
14. Repeat step 4 with each of the other screws.

Building Blocks of Engineering

Screw Discovery Lab Continued

Remember: A screw is an inclined plane wrapped around a rod.

Results:

How many ridges on the screw	How many turns it took to screw it in	How much effort on a scale of 1-10
0 (Nail)	0	

Conclusions:

Question:

What is the trade-off for using less effort?

STEM Middle School
Mechanical Engineering II Lesson # 3

Mechanical Advantage of a Screw

 Remember: A screw is an inclined plane wrapped around a rod.

Evaluate:

1. Which has a smaller pitch?
2. Which has more grooves?
3. Which has less ridges?
4. Which has more threads?
5. Which has less threads?
6. Which is made of a steeper inclined plane?
7. Which is made of a gentler sloped inclined plane?
8. Which has less MA?
9. Which has more MA?
10. Which has a greater pitch?
11. Which is made of a longer inclined plane?
12. Which is made of a shorter inclined plane?

Compose:

Draw a screw with more mechanical advantage than both of the screws above.

Infer:

Draw 2 screws that you can find in everyday objects

Building Blocks of Engineering

STEM Middle School
Mechanical Engineering II Lesson # 4

Discovery Lab for a wedge

 Remember: A screw is an inclined plane wrapped around a rod.

Problem:

Can you cut a potato with a spoon?

Hypothesis:

Materials:

Plastic spoon, Potato, and Lots of napkins

Procedure:

Attempt to cut the potato in half and then into quarters using only a plastic spoon.

Results:

Conclusions:

Questions:

What would make this task a lot easier?

How can you modify the spoon so that it can achieve the task more easily?

Building Blocks of Engineering

⬤	Design opportunity – **Build a rolling toy! Use cardboard, skewers, Styrofoam, paper, plastic –you name it!**
⬤	Research- how wheels attach to an axle and still spin
⬤	Brainstorm- write all your ideas below
⬤	Sketch out your best idea below
⬤	Materials- list your materials below, and plan a procedure
⬤	Build it
⬤	Evaluate your design and how well it fulfilled the design opportunity
Additional Info	

Building Blocks of Engineering

Mechanical Advantage of a Wheel and Axle

Remember:

MA= Diameter of the wheel/ Diameter of the axle

Complete the chart below:

MA	Diameter of the Wheel (mm)	Diameter of the Axle (mm)
5	100	20
2	30	
1		37
	66	6
	51	3
25	100	
3		40
	80	8
12	60	
22		4

Review:

MA= Effort arm/ Resistance arm

MA= Diameter of the wheel/ Diameter of the axle

MA= Resistance/Effort

Mechanical Advantage of a Wheel and Axle Word Problems

Remember:
MA= Diameter of the wheel/ Diameter of the axle

Calculate:

1. Which has greater MA — a wheel with a diameter of 50 mm and a 5 mm axle, or a wheel with a 40 mm diameter and a 2 mm axle?

2. If Joey uses a wheel with a 75 cm diameter and a 5 cm diameter axle, how much resistance can he push using a force of 3 N?

3. There is a wheel and axle with a MA of 5. For every turn the wheel makes, how many turns does the axle make?

4. What is the advantage of using a wheel that is very large in diameter, as compared to the axle? The disadvantage?

STEM Middle School
Mechanical Engineering II Lesson # 8

Building Blocks of Engineering *(vertical text, left margin)*

Lazy Susan Design and Discovery Lab

●	Design opportunity – **Help! I keep storing my spices in the cabinet, but whenever I need the chili powder it's always in the back and I make quite a mess reaching for it. Design and build a device that I can place my spices on and spin it to the correct spice.**
●	Research
●	Brainstorm- write all your ideas below
●	Sketch out your best idea below
●	Materials- list your materials below, and plan a procedure
●	Build it
●	Evaluate your design and how well it fulfilled the design opportunity
Additional Info	

Building Blocks of Engineering

Get in Gear Design and Discovery Lab

●	Design opportunity – **Use a motor and at least three gears to make a wheel and axle turn at least twice as fast as the motor is spinning.**
●	Research-
●	Brainstorm- write all your ideas below
●	Sketch out your best idea below
●	Materials- list your materials below, and plan a procedure
●	Build it
●	Evaluate your design and how well it fulfilled the design opportunity
Additional Info	

Building Blocks of Engineering

STEM Middle School
Mechanical Engineering II Lesson # 10

Gear Direction

Remember: *Gears alternate in their rotation.*

Interpret and Solve:

1. if gear number three was going clockwise, in what direction would gears 2 and 4 go?
2. if gear number three was going counterclockwise, in what direction would gears 2 and 4 go?
3. if gear number 6 was going counterclockwise, which other gears would turn in that direction?
4. if gear number 6 was going clockwise, which other gears would turn in that direction?

True or False:

1. sandwiching a gear between two of the exact same gears renders the middle gear meaningless in terms of changing **direction** of the final gear

   _____

2. sandwiching a gear between two of the exact same gears renders the middle gear meaningless in terms of changing **speed** of the final gear _____

3. sandwiching a gear between two of the exact same gears renders the middle gear meaningless in terms of changing **placement** of the final gear

   _____

Building Blocks of Engineering

Gear Direction

 Watch the DesignSquad video Tour DeBBq

Summarize:

1. What did you learn from watching this video?

2. What was the piece of the Green team's bike that was able to slow the turning of the roaster?

3. What did the purple team use as a source of energy?

4. Which team's bike did you think showed better engineering? Why?

STEM Middle School
Mechanical Engineering II Lesson # 12

Gear Ratio Lab

Problem:

How does the amount of teeth in a gear affect the speed of that gear?

Hypothesis:

Materials:

Several Different Gears

Procedure:

1. Choose one gear and place it on the table
2. Choose a second gear with half the amount of teeth as the first gear
3. mesh the gears together
4. Turn the first gear one full turn. Record how many turns the second gear makes
5. Repeat this with any two gears. Always record the number of teeth on both gears and turn the first gear one full turn.
6. You should begin to notice a system

Results:

How many teeth on first gear	How many teeth on second gear	Gear ration: if you turn the first gear once, how many times (or fractions of a single time) will the second gear turn)

Gear Ratio Lab Continued

Conclusions:

Question:

How can you increase the speed of a bike that uses a 24 toothed gear connected to a 48 toothed gear?

STEM Middle School
Mechanical Engineering II Lesson # 13

Gearing Up and Gearing Down

Remember: Gearing up is when the drive gear has more teeth than the driven gear. Gearing down is when the drive gear has less teeth than the driven gear.

Classify:

Choose gearing up or gearing down as your answer:

1. increase speed _____

2. decrease speed _____

3. small gear to large gear _____

4. large gear to small gear _____

5. drive wheel is larger _____

6. drive wheel is smaller _____

7. driven wheel is larger _____

8. driven wheel is smaller _____

9. gear ration is greater than 1 _____

10. gear ration is less than 1 _____

11. more mechanical advantage _____

12. less mechanical advantage _____

13. use it to go uphill with more force and smaller distance _____

14. use it to go faster, covering more distance with less force _____

Building Blocks of Engineering

Gear Ratio

Remember: Gear Ratio = Drive/Driven

Calculate:

Drive Wheel	Driven Wheel	Gear Ratio	Gearing up or Gearing Down
65	5		
	12	4:1	
	12	1:4	
4	40		
33	99		
99	33		
200		2:1	
		1:3	Gearing down
4	44		
44	4		
	45	1:9	
45		9:1	
77	11		
21	3		
18	32		
30		3:5	

Gear Ratio Word Problems

Remember: Gear Ratio = Drive/Driven

Develop an Understanding:

1. If a drive gear has 48 teeth, and the driven gear has 8 teeth, what is the gear ratio?

2. If three gears are meshed, and the first and last gears are the same size, does the size of the middle gear matter? Why or why not? What purpose does that middle gear serve?

3. If you have a gear with 51 teeth and another gear with 17 teeth on the same axle, what will the ratio of their speeds be?

4. What is the gear ratio of a gear system where a gear with 60 teeth is sandwiched between a drive gear of 120 teeth and a driven gear of 30 teeth? Show your work!

5. What is the advantage and what is the disadvantage of gearing down?

6. What is the advantage and what is the disadvantage of gearing up?

7. Please design a gear system in which a set of wheels turn 40 times faster than the drive gear.

8. If Jessica has a gear system with a **driven** gear with 80 teeth and a **drive** gear with a 20 teeth, how much faster does her driven wheel go?

9. If Jessica has a gear system with a **drive** gear with 80 teeth and a **driven** gear with a 20 teeth, how much faster does her driven wheel go?

Building Blocks of Engineering

The Timekeeper Design and Discovery Lab

⬤	Design opportunity –**Build a timepiece that can turn a "second hand" once a second for at least ten seconds.**
⬤	Research- gears, pendulums, weights, gear ratios
⬤	Brainstorm- write all your ideas below
⬤	Sketch out your best idea below
⬤	Materials- list your materials below, and plan a procedure
⬤	Build it
⬤	Evaluate your design and how well it fulfilled the design opportunity
Additional Info	

Pressure

 Remember: Pressure = Force/Area

Solve:

Force (N)	Area	Pressure
45	5	
	11	55
66	6	
	31	3
64		8
72	12	
	15	6

Infer and Explain:

1. Would it hurt more if a 150-pound lady stepped on your toe in flat sneakers or in high heeled shoes? Why?

2. Why aren't people crushed by the weight of the air on top of them?

3. If a cube has side of 3 cm, and a force of 81 N is being exerted over it, what is the resulting pressure per cm?

Building Blocks of Engineering

STEM Middle School
Mechanical Engineering II Lesson # 18

Bernoulli's Principle Discovery Lab

Problem:

How does air pressure affect liquid in a straw?

Hypothesis:

Materials:

Cup of liquid
Plastic straw
Empty cup

Procedure:

1. Place the straw in the cup of liquid.
2. Cover the top of the straw with your finger
3. Lift the straw out of the cup. What happens to the liquid? Is this what you expected? Write your answers in the results and observations section
4. See if you can use this method to transfer the liquid from one cup to the other.

Results/Observations

Bernoulli's Principle Discovery Lab Continued

Conclusions:

Questions:

Why does the liquid in the straw act this way?

What force is the liquid defying while your finger is covering the top of the straw?

What happens when you remove your finger? Why?

Building Blocks of Engineering

Forces acting on an Airplane

Remember: Thrust, Lift, Drag and Gravity are all forces that act on an airplane.

Label the plane with the correct vector forces

Designate the Correct Force or Forces that is...

1. keeping the plane up in the air?

2. pushing the plane forward?

3. slowing a plane down?

4. engineers designing the plane with a pointed nose?

5. engineers designing the plane with curved wings?

6. engineers making the plane as compact as possible?

7. engineers making the plane out of a lightweight material?

8. engineers adding thrusters to a plane?

9. why rocket ships need a tremendous amount of fuel to exit the atmosphere?

10. why spaceships can keep going forever in the same direction without adding more force?

Aerospace Design

	Design opportunity –**Build a paper airplane that can soar the furthest distance**
	Research- aerodynamic shapes, ways to reduce drag and increase thrust
	Brainstorm- write all your ideas below
	Sketch out your best idea below
	Materials- list your materials below, and plan a procedure
	Build it
	Evaluate your design and how well it fulfilled the design opportunity
Additional Info	

Building Blocks of Engineering (vertical, left margin)

Building Blocks of Engineering

Pneumatics Overview

Remember: Pressure= force/ area.

Develop an Understanding:

1. What does pneumatics mean?

2. Name 4 benefits of using pneumatic systems:

3. Name 3 disadvantages of using pneumatic systems:

4. What is Pascal's law?

5. How is the air pressure of an open container different from the air pressure in the room?

6. What is air pressure?

7. If you increase the amount of air in a sealed container, will that increase or decrease the pressure? Why?

8. If you increase the size of a sealed container, will that increase or decrease the pressure? Why?

9. What happens to a container of compressed air if you remove the piston?

10. How does expanding air provide energy for a system?

Summarize:

Please list 5 things you learned about how compressed air can be used to do work:

Aerospace Design

⬤	Design opportunity –**Build a mini pom-pom launcher powered by air pressure! You will have one day to design and build a first prototype and a second day to test, adjust and decorate.**
⬤	Research- aerodynamic shapes, ways to reduce drag and increase thrust
⬤	Brainstorm- write all your ideas below
⬤	Sketch out your best idea below
⬤	Materials- list your materials below, and plan a procedure
⬤	Build it
⬤	Evaluate your design and how well it fulfilled the design opportunity
Additional Info	

Building Blocks of Engineering

How to Integrate This Program:

This program was developed in order to provide a **fully-supported** STEM Curriculum to your Classroom. We understand that Teachers work hard to encourage success in their classrooms, and we have developed several additional tools to be used in conjunction with *The Teacher's Guide to The Building Blocks of Engineering* and *The Building Blocks of Engineering Student Workbook.*

Available from AdvanceSTEM.com/ stemmiddleschool@aol.com:

- ❖ Comprehensive Materials Kits – all the materials you need for all the hands-on labs and activities.
- ❖ Assessment Packets- An accelerated and remedial level test for each unit, plus study guides and quizzes.
- ❖ STEM Expansion Folders- extra activities for classrooms that want to engage in even more engineering activities.
- ❖ Annotated Teacher's Workbook- a workbook with clearly delineated answers.
- ❖ The Engineer's Sketchbook -a STEAM Initiative to help integrate the Art component into STEM.
- ❖ Professional Development- In case your teachers want some hands-on training from the pros.

Contact Us Today! We Want to Help You Integrate STEM Now!

917-838-7864/ stemmiddleschool@aol.com/advanceSTEM.com

Acknowledgements:

General Support:

G. Grant

Grant Family

Editing and Proofreading:

R. A. Weiss

Mrs. A. Weiss

Logo and Design:

I.D. Weiss

Business Consulting:

A.M. Weiss

Materials:

N. Grant

A special thank you to JFS for providing the progressive platform for the development of these materials

Made in the USA
Middletown, DE
06 September 2020

17018387R00095